Oxford International Primary

2

# Science

## Student Book

Deborah Roberts
Terry Hudson

Alan Haigh
Geraldine Shaw

Language consultants:
John McMahon
Liz McMahon

OXFORD

# OXFORD
## UNIVERSITY PRESS

Great Clarendon Street, Oxford, OX2 6DP, United Kingdom

Oxford University Press is a department of the University of Oxford. It furthers the University's objective of excellence in research, scholarship, and education by publishing worldwide. Oxford is a registered trade mark of Oxford University Press in the UK and in certain other countries.

British Library Cataloguing in Publication Data

Data available

ISBN 978-1-382006552

7 9 10 8

Paper used in the production of this book is a natural, recyclable product made from wood grown in sustainable forests. The manufacturing process conforms to the environmental regulations of the country of origin.

Printed in China by Golden Cup

**Acknowledgements**

The publisher and authors would like to thank the following for permission to use photographs and other copyright material:

**Cover:** Artwork by Blindsalida. **Photos: p6a:** Zentilia/Shutterstock; **p6b:** Sergey Peterman/Shutterstock; **p6c:** Bibiphoto/Shutterstock; **p11:** Maks Narodenko/Shutterstock; **p12-13:** Paula french/Shutterstock; **p14a:** Tomasz Klejdysz/Shutterstock; **p16b:** Svitlana Tkach/Shutterstock; **p14c:** Olga_i/Shutterstock; **p14d:** marketa1982/Shutterstock; p15a: underworld/Shutterstock; p15b: Dmitry Kalinovsky/Shutterstock; **p15c:** Kuznetsov Alexey/Shutterstock; **p15d:** papkin/Shutterstock; **p16a:** Blue Images/Corbis/Image Library; **p16b:** ifong/Shutterstock; **p17:** EcoPrint/Shutterstock; **p18a:** FatCamera/E+/Getty Images; **p18b:** Rawpixel.com/Shutterstock; **p22:** Don Mason/Corbis/Image Library; **p24a:** JackF/iStockphoto; **p24b:** David Steele/Shutterstock; **p24c:** Sergey Rusakov/Dreamstime; **p24d:** Viktoria Makarova/Fotolia; **p24e:** slowmotiongli/Shutterstock; **p24f:** Smeola/Shutterstock; **p24g:** Dennis Jacobsen/Shutterstock; **p24h:** Sergey02/Dreamstime; **p24i:** Serban Enache/Dreamstime; **p24j:** Jeanninebryan/Dreamstime; **p24k:** Photos_martYmage/iStock/Getty Images; **p25a:** Anup Shah/naturepl.com; **p25b:** David McGowen/Fotolia; **p25c:** Nature Picture Library/Alamy Stock Photo; **p25d:** Cathy Keifer/Dreamstime; **p27:** Anettphoto/Shutterstock; **p28-29:** HUSSEINALSHAFAI/Shutterstock; **p29:** amenic181/Shutterstock; **p30a:** Kletr/Shutterstock; **p30b:** Olga Myosotisphotos/Shutterstock; **p30c:** wsf-s/Fotolia; **p32a:** Kozak Sergii/Shutterstock; **p32b:** Bogdan Steblyanko/Alamy Stock Photo; **p32c:** olgaman/Shutterstock; **p32d:** BJ Flavio-Massari/Alamy Stock Photo; **p33:** Brzostowska/Shutterstock; **p35:** Bogdan Wankowicz/Shutterstock; **p36a:** Vorobyeva/Shutterstock; **p36b:** Bergamont/Shutterstock; **p36c:** Vova Shevchuk/Shutterstock;

**p36d:** Mffoto/Shutterstock; **p36e:** Valentyn Volkov/Shutterstock; **p36f:** johnfoto18/Shutterstock; **p36g:** George Dolgikh/Shutterstock; **p36h:** Brent Parker Jone/Oxford University Press ANZ; **p36i:** Brent Parker Jone/Oxford University Press ANZ; **p36j:** ArtCookStudio/Shutterstock; **p36k:** Alexandra Gl/Fotolia; **p36l:** Photolinc/Shutterstock; **p36m:** Glowimages/Getty Images; **p38:** Nigel Cattlin/Alamy Stock Photo; **p40:** Anne Spelledwithane/Shutterstock; **p41:** ER_09/Shutterstock; **p42a:** Aaron Haupt/Science Photo Library; **p42b:** Nigel Cattlin/Alamy Stock Photo; **p42c:** Theo Allofs/ Corbis/Image Library; **p44-45:** Planetary Visions Ltd/Science Photo Library; **p44:** Karen Ford Photo/Shutterstock; **p45:** Arnold O. A. Pinto/Shutterstock; **p46a:** Hypnotype/Shutterstock; **p46b:** bluehand/Shutterstock; **p46c:** korkeng/Shutterstock; **p46d:** Bragin Alexey/Shutterstock; **p46e:** Andrey Lobachev/Shutterstock; **p46f:** VICUSCHKA/Shutterstock; **p46g:** Eric Isselee/Shutterstock; **p46h:** weter 777/Shutterstock; **p46i:** Pavlo Loushkin/Shutterstock; **p46j:** Anton Starikov/Shutterstock; **p46k:** Aaron Amat/Shutterstock; **p46l:** binbeter/Shutterstock; **p48a:** Prill/iStock/Getty Images; **p48b:** Edwardje/Dreamstime; **p48C:** SCOTTCHAN/Shutterstock; **p49a:** Rich Carey/Shutterstock; **p49b:** Christian Delbert/Shutterstock; **p49c:** Steve Cordory/Shutterstock; **p52:** Patrick Poendl/Shutterstock; **p54a:** tezzstock/Shutterstock; **p54b:** Kotomiti Okuma/Shutterstock; **p55:** Kristof lauwers/Shutterstock; **p58a:** 2009fotofriends/Shutterstock; **p58b:** Raulbaldean/Shutterstock; **p59:** Favious/Shutterstock; **p60a:** Paulo Oliveira/Alamy Stock Photo; **p60b:** Kev Gregory/Shutterstock; **p61:** Monkey Business Images/Shutterstock; **p62a:** Pablo Caridad/Dreamstime; **p62b:** Shutterstock; **p62c:** AndreAnita/Shutterstock; **p62d:** Patryk Kosmider/Dreamstime; **p62e:** Kletr/Shutterstock; **p62f:** Cigdem Sean Cooper/Shutterstock; **p62g:** Anton Foltin/Shutterstock; **p62h:** Sergey Uryadnikov/Shutterstock; **p63a:** sherpa/Shutterstock; **p63b:** Fortgens Photography/Shutterstock; **p63c:** Mary_arch_Tutynina/Shutterstock; **p64-65:** Pla2na/Shutterstock; **p65a:** rangizzz/Shutterstock; **p65b:** Badboo/Dreamstime; **p67a:** Evgeny Karandaev/Shutterstock; **p67b:** RTimages/Shutterstock; **p67c:** Ingram Publishing/OUP; **p67d:** Todd Taulman/Shutterstock; **p67e:** Git/Shutterstock; **p67f:** Pavel Kudryavtsev/Dreamstime; **p68a:** Joseph Gough/Dreamstime; **p68b:** AVprophoto/Shutterstock; **p68c:** F. Moscheni/Sheltered Images/Glow Images; **p68d:** xpixel/Shutterstock; **p69a:** WENN Rights Ltd/Alamy Stock Photo; **p72a:** Thodonal88/Shutterstock; **p72b:** Jordi C/Shutterstock; **p72c:** Varderesyan Nunik/Shutterstock; **p72d:** Ullstein bild/Contributor/Getty Images; **p76a:** Vyaseleva Elena/Shutterstock; **p76b:** Vidmantas Goldberg/Shutterstock; **p78a:** photosync/Shutterstock; **p78b:** pling/shutterstock; **p78:** ARENA Creative/Shutterstock; **p79:** CREATISTA/Shutterstock; **p80:** KIM NGUYEN/Shutterstock; **p81c:** ARENA Creative/Shutterstock; **p82a:** Milleflore Images - Healthy Living/Alamy Stock Photo; **p82b:** MaraZe/Shutterstock; **p83:** Diana Taliun/Shutterstock; **p84a:** silver-john/Shutterstock; **p84b:** Alexandra Lande/Shutterstock; **p84c:** Andrew Lambert Photography/Science Photo Library; **p84d:** Imageman/Shutterstock; **p84e:** Martyn F. Chillmaid/Science Photo Library; **p85:** Iakov Kalinin/Shutterstock; **p86a:** Mint Images RF/Getty Images; **p86b:** Knorre/Shutterstock; **p86c:** Sergey Goruppa/OUP; **p86d:** tescha555/Shutterstock; **p86e:** Andrew Mayovskyy/Shutterstock; **p86f:** Sergey Goruppa/Shutterstock; **p86g:** Unkas Photo/Shutterstock; **p86h:** Mayovskyy Andrew/Shutterstock; **p86i:** Maglara/Shutterstock; **p87:** sutsaiy/Shutterstock; **p88a:** Tanya_mtv/Shutterstock; **p88b:** magicoven/Shutterstock; **p88c:** Kae Deezign/Shutterstock; **p88d:** Mike Flippo/Shutterstock; **p90-91:** aryos/iStockphoto; **p91:** pixelparticle/Shutterstock; **p96a:** Lane V. Erickson/Shutterstock; **p96b:** kryzhov/Shutterstock; **p97:** IgorZh/Shutterstock.

Artwork by Six Red Marbles and Q2A Media Services Pvt. Ltd.

Every effort has been made to contact copyright holders of material reproduced in this book. Any omissions will be rectified in subsequent printings if notice is given to the publisher.

# Contents

# How to Use this Book

This Student Book for *Oxford International Primary Science* forms part of your science lessons for this year. Your teacher will introduce the ideas through whole-class activities, then you will explore them in more detail using this book, before all coming back together to discuss what you have learned. Find out more at: www.oxfordprimary.com/international-science

## Structure of the book

This book is divided into five units plus a *Being a Good Scientist* introduction and a picture Glossary:

**Being a Good Scientist**
**Unit 1** Living and Growing
**Unit 2** Growing Plants
**Unit 3** Habitats and Food Chains
**Unit 4** Uses of Materials
**Unit 5** Day and Night
**Glossary**

Each unit covers a different strand of science. You will need a science notebook to write in and to record your investigation results and conclusions.

## Being a good scientist

To be a good scientist you need to be curious and ask questions. This section will help you think about how to develop your scientific skills to work like a scientist.

## What you will find in each unit

There are three types of lessons:

**Wow** introduces each unit's scientific ideas and key words. It tells you what you will learn in the unit and lets you discuss what you already know.

**Focused** lessons cover the scientific knowledge and skills you need to learn this year.

In **What have I learned?** you review your understanding and show your teacher what you have learned about the unit.

## What you will find in the lessons

Although each lesson is unique, they have common features:

The words on the Wow pages are included in the picture glossary at the back of the book. You can add your own notes for each word.

Gives you the key words for the lesson.

Tells you what you will learn in the lesson.

Questions to help you talk to each other and share ideas about the science you are learning and the investigations you do.

Practical and research activities to investigate and report on science topics. Sometimes your teacher will ask you to use different equipment, which is available in school. They may also ask you to carry out a test in a different way, to make sure you are safe.

**Stretch zone** Challenges you to take your learning further.

**Key idea** Summarises what you have learned.

## Additional features

**Think back** Reminds you what has been covered before.

**Science fact** Interesting and amazing science facts.

Highlights the skills needed to be a good scientist.

Important notes about how to stay safe.

## Teacher's Guide

There is a Teacher's Guide to help your teacher to work out the resources needed and to offer alternative activities and approaches.

## Workbook

At the bottom of each page in this book is a link to a Workbook, where you can record your work and get extra practice to do in your lesson or at home.

# Being a Good Scientist

Science is the study of the world around us. To be a good scientist you need to be curious and ask questions. This section will help you think about how to develop your scientific skills to work like a scientist.

Scientists look carefully at the world to explain why things happen and to guess if things may happen. Science is used to develop new technologies. It also helps us know more about health and diseases. This means we can develop medicines and machines to keep people healthy.

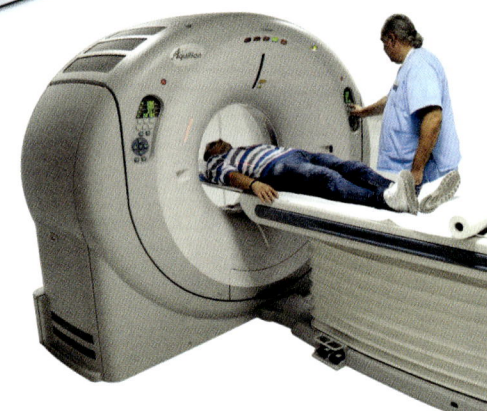

The diagram shows the steps you can take to find out about things (investigate) like a scientist.

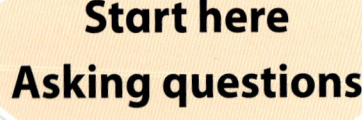
Start here
Asking questions

I think that ...

I am going to ...

I am looking for ...

I have found that ...

This means that ...

Learning to be a scientist allows you to develop scientific skills such as observing (looking), measuring and recording. It helps you to notice patterns in the things you observe and to sort things into groups. It also helps you to test our own ideas about how the world works.

## Asking questions

Scientists ask questions about the world around them. This is called scientific enquiry.

A good way to start is to think of questions that start with words such as 'how', 'which', 'what', 'do' and 'does'.

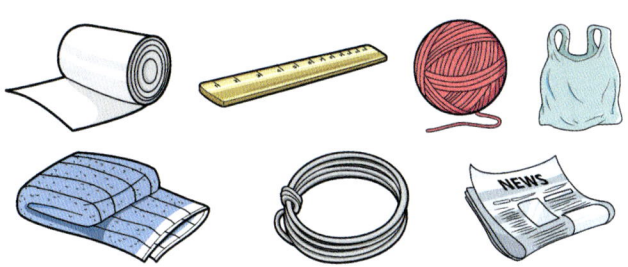

How are these materials different?

Which materials will feel soft?

Think of your own questions to ask about the different materials.

The questions you ask will give you a good start to your investigation.

## I think that ...

Next, scientists try to work out or guess what will happen. Scientists call this a prediction.

They need to talk about their ideas and what they think will happen.

You might have already learned something about the question you are trying to answer. Scientists usually know something before they make predictions.

Use what you know about materials to help you think about this question.

Which of the materials will stretch the most?

Do you think the plastic bag will be very stretchy?

What did you think about to help you choose?

## I am going to ...

Scientists plan what they are going to do. They always discuss their plans before they start. This helps to check that the plan will work.

Scientists make their investigations fair by following some simple rules:

- They think about what they will keep the same.

- They think about what they will change.

For example, when investigating which substances dissolve in water, you should use the same volume of water to test each one. You should also stir them for the same amount of time, and have them at the same temperature. This makes sure that the only change is the substance you are adding to the water.

Scientists think about the equipment they need. They make a list and make sure everything is available.

For example, if you are going to test substances dissolving, you might make an equipment list like this:

small dishes
pen and paper
measuring jug
substances
spoon
water

### Science fact

Scientists do not always plan their own investigations. Sometimes they follow other scientist's plans. This is why it is very important to make the plans easy to follow.

## I am looking for ...

Scientists look closely at what is happening in their investigation. They use all of their senses. These are called observation skills.

During an investigation you will look, listen, smell, touch and sometimes taste.

**Warning!** Only smell, taste and touch things if your teacher tells you it is safe. Many things can be poisonous.

You may need to use equipment to help with your observations. Some of the pieces of equipment you will use in this book are shown below.

Good scientists use equipment carefully. They take a measurement more than once. This is to make sure they have not made any mistakes.

You can practise measuring the heights of three objects to find out how many centimetres high they are.

Which piece of equipment will you use?

When measuring the plant where will you start from and where will you end?

Tell your partner your measurements.

## I have found that ...

Scientists write down or record what they have found from their observations and measurements. This helps them to see patterns or to sort things into groups.

There are lots of different ways to record results.

### Table

One way is to complete a table.

You could use a table like this to record litter found in a survey:

| Type of litter | Number of items |
|---|---|
| plastic | 8 |
| paper | 12 |
| cardboard | 4 |
| metal | 2 |
| rubber | 3 |
| glass | 6 |

What was the most common type of litter?

What was the least common type of litter?

Now you can design your own table.

Ask the students in your class about their favourite exercise.

Design a table for your results and record them.

### Charts

Results from tables can be shown as charts or graphs.

This chart shows the results of the investigation about litter found in a local area.

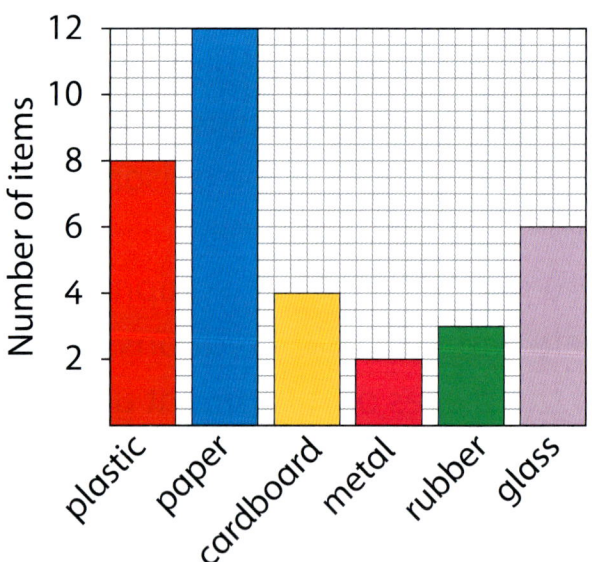

The type of litter is plotted along the bottom of this chart.

The number of items found are plotted from bottom to top.

Using charts can make it easier to see patterns in results.

## Drawings and models

Scientists may draw or make models of the things they are studying. Models help scientists to see and explain how things work. Science drawings are not like the pictures you paint. Scientific drawings are much simpler.

You might need to draw a plant or an animal, or a piece of equipment.

This is a simple drawing of an orange but you can still tell what it is.

## Photographs and videos

Scientists might also take photographs and video clips of their investigations and results.

This is a very accurate way to record results.

## This means that ...

The last stage of an investigation is when scientists looks at their results carefully.

They work out if the results have helped them to answer their investigation question.

The questions they might ask are:

Can I *see* any patterns?

Are any results unusual?

Was my prediction correct?

Could I have done anything *better*?

Finally, they will always think about how to make their investigation better.

It is useful to fill out an investigation planning form. This sets out all the stages of your investigation. It helps you to remember everything you need to think about. Your teacher can give you one of these.

# 1 Living and Growing

In this unit you will:

- find out that animals and humans need water, food and air
- find out why we need a healthy diet
- explore why exercise is important
- learn about good hygiene
- notice that babies grow into adults
- explore some animals and their offspring.

Look closely at the photograph of the waterhole.
Point out the different animals.
Discuss what the animals are doing.

adult   diet   exercise
grow   hygiene
movement   offspring
parent   teenager
toddler

**Science fact**

Scientists have proven that regular exercise helps to keep people happy.

What are the children doing?

Why is this good for them?

Where are they getting their energy from?

■ For more activities, go to Workbook 2 pages 12–13.

# What animals and humans need to live

In this lesson you will find out what animals and humans need to live.

**Key words**
breathe
drink
eat
grow
move
reproduce

Animals and humans need to be able to breathe, eat and drink to stay alive. These are the processes for life.

Look at the photographs.

Name any life processes you can see.

Predict what would happen if these life processes stopped.

Animals need to find food and water. Some animals eat other animals to get their food. Animals have to breathe air to stay alive.

Eating food and drinking makes animals grow into adults, so they can reproduce and have young.

Moving, growing and reproducing are also life processes.

## Science fact

The cheetah is the fastest land mammal. It can go from standing to 96 kilometres per hour in three seconds.

14

■ For more activities, go to Workbook 2 page 14.

Look at the photographs of the animals and their food in the word boxes. Discuss with a partner which food each animal will eat.

| eggs | grass | small insects | fish |

Now agree on one word to describe how each of the animals moves.

## Food survey

You are going to plan and carry out a survey of foods that humans eat.

1  Ask the students and teachers in your class what they ate for their last meal.

2  Design and complete a table to record your findings.

3  Include whether each food is from plants or animals.

Share your findings with the class.

## Be a scientist

When scientists carry out surveys they record the results straight away. They often use tables.

▶ page 10

## Key idea

Animals and humans must breathe, eat and drink to stay alive.

## Stretch zone

Plan how you would investigate local animals to find out what they eat.

■ For more activities, go to Workbook 2 page 15.

# Eating and drinking

In this lesson you will find out why we need a healthy diet.

**Key words**
diet
drink
eat
energy
food
healthy/
unhealthy
water

What we eat and drink every day is called our diet. We need to eat five kinds of food to stay healthy.

This plate shows the types of food needed for a healthy diet.

What are the family doing? Why is this important?

bread and cereal

fruit and vegetables

protein

milk and dairy

fat and sugar

Think about what you had to eat yesterday. Make a poster to show the different food you ate.

16

Some foods give us the energy we need to stay active.

All animals and humans also need to drink lots of water to stay healthy.

We cannot eat just one or two kinds of food. This will make us unhealthy.

## Planning a healthy meal

You are going to plan a healthy meal with your group.

Use the photograph of the food plate to help you.

You can also use the internet and magazines to collect pictures of different food.

1 Decide what food and drink you will have in your meal.

2 Make a menu showing your meal.

Starter: Chorba soup with bread;

Main: Chicken and rice with okra;

Dessert: Fruit salad with yoghurt

**Stretch zone**

Swap your menu with a partner. Do you think their meal is healthy? Explain your thinking.

**Key idea**

The right sort of food and drink keeps us healthy.

■ For more activities, go to Workbook 2 page 17.

# Exercise is important

In this lesson you will explore the importance of exercise to humans.

**Key words**
energy
exercise
heart rate
muscle
pulse

Which photograph shows people using up the most food for energy?

What happens if food is not used for energy?

When animals and humans exercise they use up energy foods, such as sugars and fats. This gives them the energy needed to move muscles. This makes the muscles stronger.

If energy foods are not used up, they are stored as body fat. If an animal is very overweight, with a lot of body fat, they are called obese.

The heart is a muscle. It grows stronger with exercise. You can feel your heart beating at different points on your body. This is called your pulse.

**Science fact**

As the heart becomes stronger it beats less. Fit people have a slower pulse rate. Top class cyclists have a pulse rate lower than 40 beats per minute.

How to take a pulse at your wrist.

A pulse is counted for 60 seconds. This gives the heart rate per minute.

18

## Measuring heart rate

Work with a partner.

1 Take your partner's pulse before they do any exercise.

This is their resting heart rate. Record it.

2 Ask them to run gently on the spot for 30 seconds.

Take their pulse again. Record it.

3 Now ask your partner to measure your pulse rate before and after exercise.

4 How did your pulse rates change? Did you notice any patterns in your results?

Talk about the different exercises in this picture.

As well as fitness and lowering body fat, what else can exercise do for you?

### Key idea

Regular exercise keeps people fit and healthy. It also stops them from becoming overweight or obese.

**Stretch zone**

Design a poster to persuade people to take more exercise. Use facts and pictures.

■ For more activities, go to Workbook 2 page 19.

# Good hygiene

In this lesson you will describe the importance of good hygiene.

**Key words**
hygiene
microbes
washing

What are these children doing?

Talk about how often they should do this. Why is it important?

Hygiene is about how we keep our bodies, our clothes and our homes clean.

This stops tiny living things such as bacteria and viruses from damaging our bodies and teeth or making us ill. These living things are called microbes.

## Hand-washing test

Your teacher will divide your class into four groups, A–D.

**A** No hand washing

**B** Washing with cold water

**C** Washing with warm water

**D** Washing with warm water and soap

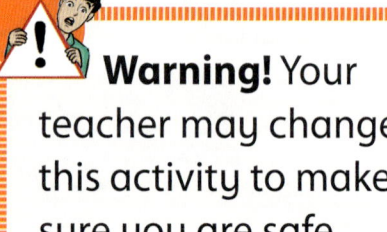

**Warning!** Your teacher may change this activity to make sure you are safe.

**1** One person in your group will rub cooking oil and sand into their hands. They will then wash their hands if in Group B–D.

**2** Next, they shake hands with one person in the group.

■ For more activities, go to Workbook 2 page 20.

**3** This person then shakes hands with another person in the group.

**4** Keep doing this until everyone in your group has shaken hands.

**5** Inspect all of the hands in your group. Inspect all of the hands in the other groups.

**6** With your group, research how to properly wash hands and make a guidance poster for the class.

What does this investigation tell you about washing hands?

There can be up to ten million microbes on your hands.

You must wash your hands after using the bathroom.

You must wash your hands before touching food.

You must wash your hands after touching animals and rubbish.

You must wash your hands if you blow your nose or cough or sneeze.

 **Warning!** When you sneeze, blow your nose or cough you should always try to use a tissue and then throw it away. This stops the spread of microbes.

**Key idea**

Good hygiene is very important. It stops us getting or passing on diseases.

Look at the pictures. Talk about how these people are keeping themselves healthy. What would happen if people did not do these things?

■ For more activities, go to Workbook 2 page 21.

# Families

In this lesson you will learn that babies grow into adults.

## Key words
adult
baby
child
family
reproduction
teenager
toddler

From child to adult

Look at the family.

How many people are in the family? How many are adults? How many are children?

Humans have babies. Having babies is called reproduction.

Babies start small and then grow. They become toddlers, teenagers and then adults.

Human life stages

Human babies need lots of help to grow and stay healthy.

What do babies need to help them grow and stay healthy?

22

■ For more activities, go to Workbook 2 page 22.

## Measuring heights

Your teacher will put you into groups. Do you think all the people in your group are the same height?

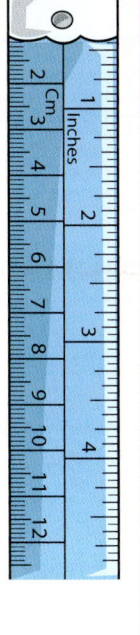

**1** Write your prediction in your notebook like this. Choose the correct words from the box to complete your prediction.

I think all the people in my group will be _____.

> the same height    different heights

**2** The picture shows how to measure a person's height. Measure the height of everyone in your group.

**3** Record your results.

How tall was the tallest person?

How tall was the shortest person?

**4** How did you make your measurements as accurate as possible?

Discuss your results with your group. Compare your results with your prediction.

### Key idea

Some humans have babies. Babies grow into toddlers, then teenagers, and finally adults.

■ For more activities, go to Workbook 2 page 23.

# Growing up

In this lesson you will discover that babies are called offspring and that offspring grow into adult animals.

**Key words**
adult
baby
offspring
parent

Animals make babies. We call the babies their offspring.

Point to the parent for each offspring. The first one is shown with a red line.

Look at the photograph of the tigers.

What is the adult tiger doing?

Can the young tiger look after itself?

How does the adult tiger help the young tiger to stay alive?

24

Not all offspring look like the adult parents.

Look at these photographs of families. What are the differences between the adult and the offspring?

## Researching adults and their offspring

1 Choose one of the animal cycles below.

egg → chick → chicken

egg → caterpillar → pupa → butterfly

spawn → tadpole → frog

lamb → sheep

2 Find out about how the offspring develop into adults.

3 Make a paper plate presentation. Use one plate for each stage of the animal's life. On the last plate, draw what the adult looks like or use pictures from the internet.

4 Make a classroom display of all your plates.

Check how much you know.
Try the puzzles on pages 26–27.

### Key idea

Animals make babies, which are called offspring. The offspring grow into adult animals.

■ For more activities, go to Workbook 2 page 25.

# What have I learned about living and growing?

**1** Order these four life stages, 1 to 4. Write the number under each picture.

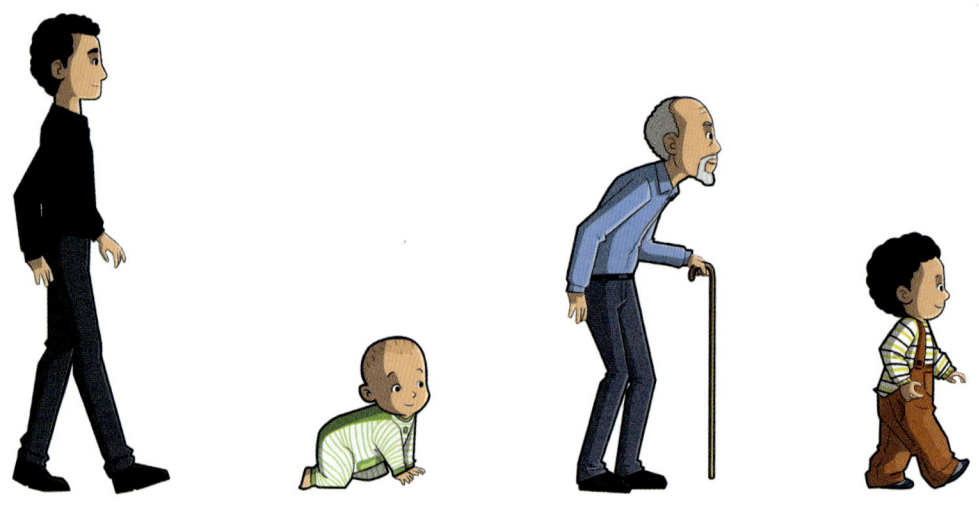

_____   _____   _____   _____

**2** Tick the hygiene action we should always do before touching or eating food.

■ For more activities, go to Workbook 2 page 26.

**3** Which of these foods do we need most of in our diet? Tick three.

**4** Which of these is a good reason to take exercise? Underline your choice.

It keeps us busy. It keeps us healthy. It saves money.

**5** There is an insect called a mayfly which lives for only one or two days. Why is it important that it reproduces in its short lifetime? Underline the correct box.

To collect enough food.

To get rid of eggs.

To make offspring so mayflies do not die out.

**6** Imagine you are alone in the desert. What do you need to stay alive? Circle all the correct words.

books    clothes    food    soap    water

■ For more activities, go to Workbook 2 page 27.

In this unit you will:

- explore how seeds and bulbs grow into plants
- discover that plants need light, water and the right temperature to grow.

Look at the photograph.

Which of these fruits and vegetables have you seen before?

How many can you name?

bulb  germination
grow  light  plant
seed  temperature
water

**Science fact**

Scientists, farmers and gardeners have developed many ways to grow new plants. They use seeds but also cuttings and even pieces of leaf grown in a laboratory.

Look at the photograph. What is the person doing? Discuss how the seeds might develop.

■ For more activities, go to Workbook 2 pages 28–29.

# Growing plants

In this lesson you will explore how seeds grow into mature plants.

## Key words

plant
seed
seedling

## Think back

Can you name the parts of a plant?

Look at these photographs.

A sunflower seedling

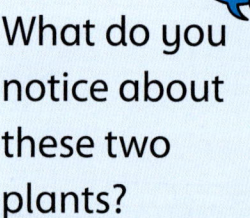

A fully grown sunflower

What do you notice about these two plants?

Name any parts of the plants that you see.

How are the plants similar? How are the plants different?

## Measuring plants

1 Use the rulers next to each photograph above to measure the height of each plant.

2 Write down the height of the seedling.

3 Write down the height of the fully grown sunflower.

The plants in the photographs are sunflowers.

Sunflowers grow very tall. They grow from seeds that look like this.

## Be a scientist

Good scientists use equipment carefully. They take a measurement more than once. This is to make sure they get an accurate result.

▶ page 9

■ For more activities, go to Workbook 2 page 30.

## How to grow sunflowers

You are going to grow some sunflowers.

1 Get two pots. Put some pebbles in the bottom of the pots.

2 Fill the pots with compost.

3 Plant two sunflower seeds in each pot. Add some water.

**Warning!**
Wash your hands after planting the seeds.

4 Check the pots every day. Keep the pots damp but not too wet.

How many days pass before you see seedlings?

5 Measure the height of your seedlings every day for two weeks.

How tall does the tallest one grow?

### Be a scientist

Scientists often have to time how long things take in an experiment. They use a timer or an accurate clock.

▶ page 9

### Key idea

We can help seeds to grow if we look after them.

■ For more activities, go to Workbook 2 page 31.

# Plants from seeds and bulbs

In this lesson you will find out why plants make seeds and bulbs.

**Key words**
bulb
grow

Look at these photographs.

How are the fully grown plants the same?

How are the plants different?

How are the seeds the same?

How are the seeds different?

## Plant and seed survey

Your teacher will take you outside to a park or the school grounds.

1 Look at the plants in the area. This includes trees.

Find as many different types of seeds as you can.

Collect some examples.

**Warning!** Do not pick seeds up unless your teacher tells you they are safe.

Were the seeds close to the parent plant or far away?

2 Back in the classroom you can use books and the internet to find the names of your seeds.

3 Make a large poster to display your seeds.

**Be a scientist**

Scientists would use a ruler or a measuring tape to find out exactly how far away from the parent plant the seeds were.

▶ page 9

32

■ For more activities, go to Workbook 2 page 32.

## Why do plants make seeds?

Flowering plants and trees make seeds.

You have seen that seeds can grow into new flowering plants.

This helps this type of plant to spread to new areas.

these seeds land near the trees

seeds spreading in the wind

### Acting as a plant

Pretend you are a flowering plant.

1 Fold up some pieces of paper. These are your seeds.

2 Stand in one place. Throw your seeds. Try to spread your seeds as far away from you as possible.

Why do plants need their seeds to spread far away?

Some plants make bulbs. These are swollen food stores that help the plant to survive cold conditions.

Bulbs can be planted and they will grow into new plants.

### Stretch zone

Write a plan on how you would investigate the growth of three different bulbs.

**Key idea**

Flowering plants produce seeds and bulbs to make new plants.

■ For more activities, go to Workbook 2 page 33.

# Measuring plants

In this lesson you will observe and measure growing plants.

**Key words**
germination
root
shoot

### Think back

Flowering plants produce seeds to make new plants.

Seeds and bulbs contain a food store. The plant uses this as it starts to grow.

When a seed starts to grow it is called germination.

Once a plant has leaves, it can then make its own food.

## Do all plants grow at the same speed?

You are going to compare the growth of two different seeds.

You will find out if they make new plants and grow at the same speed.

1 Plant the seeds in different pots. Add three of the same seeds to each pot.

2 Water the pots and leave them on a windowsill.

3 Check your pots every three days, for two weeks.

4 Keep the pots damp but not too wet.

5 Measure any seedlings that appear.

6 Use a table like this to record your results. What do your results tell you?

| Name of seed | Height of plant in centimetres (cm) | | | | |
|---|---|---|---|---|---|
| | After 3 days | After 6 days | After 9 days | After 12 days | After 14 days |
| | | | | | |
| | | | | | |

34

■ For more activities, go to Workbook 2 page 34.

Scientists think that germination in seeds starts when they send out a small new root and a small new shoot.

young shoot

seed

young root

## Investigating germination

You can test to see if seeds grow as scientists think.

1 Put some damp paper towel around the inside of a glass jar.

2 Place a seed between the paper and the glass. Keep the paper damp but not too wet.

3 Leave your jar on a windowsill.

4 Look at your seed every day for two weeks. Draw a picture of what it looks like.

5 Compare your findings with the rest of your class. Did the seeds grow as the scientists said?

**Key idea**

To find out how well plants are growing, we can observe and measure them.

■ For more activities, go to Workbook 2 page 35.

# Fruits and vegetables

In this lesson you will explore some plants that we can eat as food.

**Key words**
fruit
vegetable

We grow some plants because they provide food to eat.

Can you name these fruits, vegetables and seeds?

Talk about which you have eaten.

Humans and animals eat the food provided by plants.

We also grow some plants because they look pretty and smell sweet.

36

■ For more activities, go to Workbook 2 page 36.

## Survey of fruits and vegetables

1. You will visit a market or local shop.

2. Look for any examples of fruits, vegetables, seeds and bulbs.

3. Draw or take photographs of any that you find.

4. Do some research to name them.

5. Make a display of your photographs to tell people about your visit.

**Stretch zone**

Choose one of the fruits, vegetables, seeds or bulbs from your shop visit. Find out more about it.

How is it eaten? Has it got any other uses? How is it grown? Present your findings to the class.

**Key idea**

Many of the fruits and vegetables we eat have grown from seeds or bulbs.

**2 Growing Plants**

■ For more activities, go to Workbook 2 page 37.

# Do plants need light?

In this lesson you will find out that plants need light to grow.

## Key words
dark/light
grow
healthy/unhealthy

Which plant looks healthy?

Which plant looks unhealthy?

What clues did you use to help you to decide?

All living things need help to grow. Humans need food, warmth, water and air for example. Plants also need the right things to grow.

**Be a scientist**

Scientists only change one thing in an investigation. They try to keep everything else the same so the test is fair.

▶ page 8

## Do plants need light to grow?

You will investigate whether plants need light to grow. You need two containers to grow cress seeds.

1 Label one of your containers 'light'. Label the other one 'dark'.

2 Put some cotton wool in the bottom of each container.

3 Add a little water and put the seeds on top.

38

■ For more activities, go to Workbook 2 page 38.

**4** Put one container in a light place.
Put the other container in a dark place. Keep the cotton wool damp but not wet.

**5** Discuss how you can make your investigation fair.

**6** Predict which seedlings will grow the best.

light

dark

**7** What do the plants grown in the light look like? Draw them in your notebook.

**8** What do the plants grown in the dark look like? Draw them in your notebook.

**9** Was your prediction correct? Think about your results to answer the question.

 **Stretch zone**

Do you think seeds and bulbs need light to grow? Explain your thoughts to a partner.

**Science fact**

Green plants use energy from the Sun to help them make food. This process takes place in the leaves.

**Key idea**

Plants do not grow well in the dark.

# Do plants need water and warmth?

In this lesson you will find out that plants need water and the right temperature to grow.

**Key words**
temperature
water
wilt

The owner of these plants remembered to check the plants for insects but what did they forget?

If a plant does not have water it gets soft and limp. Scientists say the plant has 'wilted'.

Look at the photograph. Do you think these plants have been looked after?

Discuss the clues you used to answer the question.

**Science fact**

Too much water can be bad for plants. This is why droughts (lack of water) and floods (too much water) can damage plants.

You can design an investigation to show that plants do need water.

40

■ For more activities, go to Workbook 2 page 40.

## Do plants need water to grow?

1  You will be given two identical plants in pots.

   Draw both plants in your notebook at the start of the investigation.

2  Label one pot 'water' and the other pot 'no water'.

3  Put the pots in the same place. A sunny windowsill is perfect.

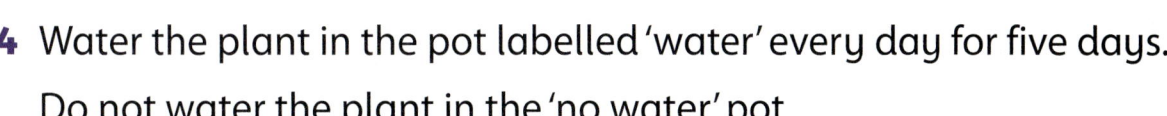

4  Water the plant in the pot labelled 'water' every day for five days.

   Do not water the plant in the 'no water' pot.

5  Look at the plants carefully every day.

   Do this for five days.

6  Draw both plants in your notebook at the end of the investigation.

7  What do your results tell you? Did you notice any patterns in your results?

We need to care for plants to help them grow. Plants need water and the right temperature.

### Stretch zone

Plan an investigation to find out if plants grow better in the cold or in warmth.

What do you predict?

**Key idea**

If a plant does not get enough water it will wilt and die.

Check how much you know.
Try the puzzles on pages 42–43.

■ For more activities, go to Workbook 2 page 41.

# What have I learned about growing plants?

**1** For each of the stages 1–4 below, draw a line to the correct picture to show how to grow sunflowers.

| **1** Get a pot. Put some pebbles in it. | **2** Fill the pot with compost. | **3** Plant and water two sunflower seeds. | **4** Water the seedlings. |
|---|---|---|---|

**2** Look at the photographs of plants and read the sentences. Circle the correct words to complete each sentence.

**a** This plant **has / does not have** enough water.

**b** This plant **has / does not have** enough water.

**c** This plant has been grown in the **light / dark**.

■ For more activities, go to Workbook 2 page 42.

**3** Which is true for a plant grown in warm conditions and then moved to cold conditions? Tick the correct answer.

It will grow well. ☐

It will grow better than before. ☐

It will not grow well in the cold. ☐

**4** Complete the missing words in the sentences below.

Many plants grow from a small _s_ ___ ___ ___.

Plants can grow from swollen food stores called _b_ ___ ___ ___ ___.

We can plant seeds in a pot full of _c_ ___ ___ ___ ___ ___.

The main things that plants need to grow are _w_ ___ ___ ___ ___

and _l_ ___ ___ ___ ___.

A plant that does not have enough _l_ ___ ___ ___ ___ will look

_y_ ___ ___ ___ ___.

A plant that does not have enough _w_ ___ ___ ___ ___ will

_w_ ___ ___ ___.

**5** A student planted some seeds in compost in pots. Tick all the advice you would give them to help their seeds germinate. There is more than one answer.

Leave the pots in sunlight. ☐

Water the pots regularly. ☐

Keep the pots warm. ☐

Don't water the pots. ☐

Keep the pots cold. ☐

■ For more activities, go to Workbook 2 page 43.

# 3 Habitats and Food Chains

In this unit you will:

- discover living and non-living things
- explore different animals and plants, and how they are suited to where they live
- identify different habitats
- explore how animals get their food from plants and other animals
- understand ways to care for the environment.

Look at the large picture. With a partner, talk about what the picture shows.

Decide why some parts are blue, some parts are green and other parts are brown.

adapted    environment
food chain    habitat
living    micro-habitat
minibeasts    non-living
pollution

Part of the Western Desert of Egypt did not have a drop of rain for 17 years!

Look at the photographs of the polar bear and the tree.

With a partner, talk about what would happen to these living things if the tree moved to the Arctic and the polar bear moved to the desert.

**45**

■ For more activities, go to Workbook 2 pages 44–45.

# Living or non-living?

In this lesson you will learn that animals and plants are living things and that some things have never been alive.

**Key words**
alive
living/non-living

### Think back

Think back to what you have learned about animals and plants so far.

How do we know if something is living or has never been alive?

Discuss the photographs. Decide which show living things and which show non-living things. Tell your partner which things are living.

■ For more activities, go to Workbook 2 page 46.

Living things can breathe, grow, move and eat.
Non-living things cannot do any of these things.

Discuss the words in the word bank. Point to
the ones that describe what living things can do.

> tiger eat grow wood breathe glass
> chair plastic water move table stone walk

Look at the sentences below.

Now read the words in the word box below.

Talk about the words then write out each of the
four words on separate cards. Then take it in turns
to read the sentence. Hold up the correct word
when you reach the gap.

Plants and ___ ___ ___ ___ ___ ___ ___ are
living things.

Living things can grow, ___ ___ ___ ___ and eat.

Buildings and ___ ___ ___ ___ ___ have never
been alive.

Things that are not alive cannot ___ ___ ___ ___,
move or eat.

> animals   grow   move   rocks

**Stretch zone**

Is a piece of wood alive? Explain your answer to
a partner.

**Key idea**

We can group
everything in
the world into
living or non-
living things.

■ For more activities, go to Workbook 2 page 47.

# Where do animals and plants live?

In this lesson you will find out about different animals and plants and where they live.

**Key words**
animal
food
plant
shelter
sunlight
water

What can you see in the picture that helps people to live and keep safe?

Look at the photographs below.

Why do antelopes live here?

Why do palm trees live here?

Why do antelopes and palm trees not live here?

■ For more activities, go to Workbook 2 page 48.

Animals and plants need water, food and shelter.
Most plants also need sunlight.

Discuss where each of these animals lives. Match the numbers to the letters. The bird lives in the nest, so 1 and B go together.

Which place do you think is the best place for the dolphin to live? Explain why you think this.

**Stretch zone**

Write two sentences to explain why there are very few plants in dark caves.

**Key idea**

Different animals and plants live in different places.

49

■ For more activities, go to Workbook 2 page 49.

# Living in different environments

In this lesson you will identify how some environments are the same and how they are different.

**Key words**
environment
minibeasts

Think about the animals and plants living in your region. Do they all live in the same type of environment?

There are many different types of environments in the world.

Some are very dry and others are very wet.

Some are very cold and others are very warm.

The animals and plants in an environment are able to survive there and find food and water.

Look at the four different environments.

How are they different?

How are they the same?

For each one, decide how the animals and plants are suited to the environment.

■ For more activities, go to Workbook 2 page 50.

Small animals such as insects can be called 'minibeasts'.

Scientists sometimes catch minibeasts with traps.

## Finding minibeasts

Choose a local environment to study.

The top of the container is just above the soil.

1  Set up your minibeast trap.

2  Leave the trap overnight.

3  In the morning, look inside and count the number of minibeasts you caught.

**Warning!** Never touch minibeasts as they could bite or sting.

4  Take a photograph of them inside the container. You could use a book or the internet to try to identify some of them.

5  Free the minibeasts back into the environment as soon as you can.

6  Design and make a minibeast scrapbook.

### Be a scientist

A good scientist would set up many traps over many days to get a more accurate idea about the minibeasts in an area.

▶ page 9

### Key idea

Environments can be different but they all give water, air and food to the animals and plants that live there.

### Stretch zone

Predict a place near you where you would find a) a lot of minibeasts and b) only a few or no minibeasts. Use traps to test your predictions.

**3** Habitats and Food Chains

**51**

# Comparing habitats

In this lesson you will learn that the type of environment affects the types of animals and plants that live there.

**Key words**
adapted
habitat
micro-habitat

Shady oasis

An environment, such as a desert or a mountain, is made up of smaller places where animals and plants live.

The place a plant or animal lives is called its habitat. This is its natural home.

Habitats can be different sizes. Some are large, such as lakes or grasslands. A very small habitat is called a micro-habitat, such as where insects live under stones or in leaf litter.

Look at the photograph of the oasis in the desert.

Can you identify some different habitats within the desert environment?

### Stretch zone

Plan how you could find out which habitat in the desert environment has the most minibeasts.

When an animal is suited to its habitat, we say it has adapted to live there.

For example, fish have gills to breathe underwater and fins to help them swim. A polar bear has thick fur to keep it warm.

■ For more activities, go to Workbook 2 page 52.

Plants are also adapted to their habitats. A lot of water escapes from plants, especially through the leaves.

small, rolled-up leaves

thick, waxy coating

Look at the picture of the cactus.

How do its two adaptations stop it from losing water?

## Water loss from leaves

You are going to investigate how plants stop water being lost from their leaves.

1 Wet three pieces of cloth. Copy the method in the pictures.

Wet cloth in sealed plastic bag

Open wet cloth

Rolled wet cloth

2 Put the cloths outside in the Sun. Predict which piece of cloth will stay wet for the longest time.

3 Observe the pieces of cloth until the first cloth is dry. Check to see how wet the other cloths are.

4 Record your findings in your notebook.

5 What does your investigation tell you about why the cactus has waxy surfaces and rolled up leaves?

## Key idea

Animals and plants are adapted to their habitats to survive.

**3 Habitats and Food Chains**

■ For more activities, go to Workbook 2 page 53.

# Animals and their habitats

In this lesson you will explore how the type of environment affects animals.

**Key word**
habitat

Animals live in different types of environment. The ones that are adapted better will grow and survive.

tough mouth for eating spiky plants

long eyelashes to keep the sand out

hump for storing food

long legs

hair to keep the Sun off

padded feet for walking on sand

With a partner, talk about how the camel is adapted to the desert but not to cold, icy places.

Tell your partner why the camel and the polar bear are not adapted to live on a coral reef.

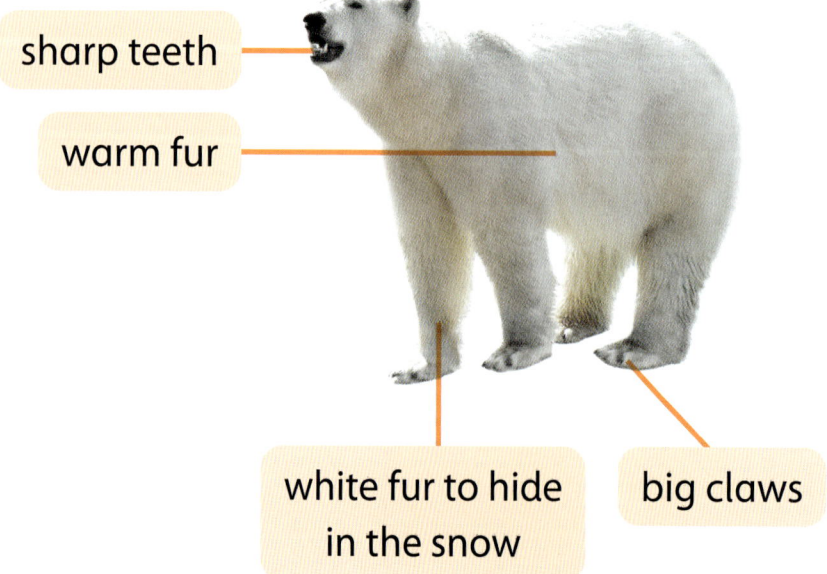

sharp teeth

warm fur

white fur to hide in the snow

big claws

## Making a minibeast hotel

Minibeasts will find food and shelter in pieces of wood and other materials.

1 Design your own minibeast hotel.

This can be a simple collection of logs or twigs.

You could make something larger using sticks, old pots and even recycled drinking straws.

A minibeast hotel

2 Check your minibeast hotel to see what is using it as a habitat.

You could use your minibeast trap, or your teacher may give you a pooter.

**Warning!** Make sure you cannot swallow any minibeast you suck up into the pooter.

Using a pooter

3 Identify at least three micro-habitats in your minibeast hotel and record what is living in them.

**Key idea**

Animal's bodies are adapted to the type of habitat they live in.

**Stretch zone**

Research why it is important to have a lot of minibeasts in an environment. Clue: What might eat them, and what might they eat?

■ For more activities, go to Workbook 2 page 55.

# Simple food chains

In this lesson you will explore how animals get their food from plants and other animals.

**Key words**

carnivore
consumer
energy
food chain
herbivore
omnivore
producer

**Think back**

What can you remember about what eats what?

What does the bird want to do? Why?

Plants can make their own food. They use energy from the Sun to help them do this.

Animals cannot make their own food. They are consumers. They have to eat, or consume, plants or other animals.

Feeding relationships can be shown as a food chain.

The arrows show how energy and food is passed along the chain.

Plants are called producers because they can produce their own food.

Animals that eat plants are called herbivores.

Animals that eat other animals are called carnivores.

Animals that eat plants and animals are called omnivores.

Talk about what you have eaten today. Are you a herbivore, a carnivore or an omnivore?

56

■ For more activities, go to Workbook 2 page 56.

# Identifying food chains

1 Research:
- three plants that live in your area
- three animals that eat the plants
- three animals that eat other animals.

Download or draw some pictures of each one.

2 Draw three circles on a large sheet of paper.

Label one of the circles 'producers', one 'herbivores' and one 'carnivores.'

3 Sort out your pictures so that each living thing fits into the correct circle.

producers    herbivores    carnivores

4 Take it in turns to take a picture of a living thing from each of the three circles and arrange them into a food chain.

5 Choose two of your food chains and present them as a poster.

 **Stretch zone**

Draw two food chains that include you.

**Science fact**

Food chains can be made up of four or five living things. If one thing is removed from the chain, it can be a disaster for the other animals.

**Key idea**

Food chains show feeding relationships.

■ For more activities, go to Workbook 2 page 57.

# Damage to habitats

In this lesson you will explore ways habitats can be harmed.

Environments can be damaged by natural disasters such as earthquakes, volcanoes and hurricanes.

Human activities such as building, making reservoirs, mining and clearing forests can also damage environments.

**Key words**
damage
deforestation
natural disaster
protect

Compare the two photographs. Which river do you predict will have the most animals and plants?

Which river is a damaged environment?

**Stretch zone**

Think about the ways you might be helping to damage habitats with the way you live. What can you do to protect habitats? Make a list.

58

■ For more activities, go to Workbook 2 page 58.

Clearing trees from a forest is called deforestation.

As well as removing the habitats for millions of animals and plants, deforestation can cause other damage.

## How does deforestation cause damage?

1 Set up two sand trays, like the ones in the picture.

Make the slope the same in each.

2 Add model trees to one – you could use twigs or drinking straws.

Push the trees deeply into the sand.

3 Predict what will happen when water runs down both slopes.

4 Test your prediction by carefully trickling water down each slope.

Observe what happens. You could film it.

Was your prediction correct?

5 Present your findings to the class. You could demonstrate your model.

6 Use your findings to explain why deforestation can cause damage to an environment and its habitats.

**Stretch zone**

Find out where forests are being cleared throughout the world.

List some of the reasons why this is happening.

### Key idea

Habitats can be harmed by natural disasters and by human activities.

■ For more activities, go to Workbook 2 page 59.

# Protecting the environment

In this lesson you will learn about ways to care for the environment.

**Key words**
litter
pollution

With a partner, talk about the types of pollution you can see.

Some substances that are put into the environment can be poisonous and harmful. This is called pollution.

We share our environment with plants and animals, so we need to take care of it.

**Science fact**

More than 12 million tonnes of plastic ends up in the oceans every year. This includes plastic bottles and shopping bags.

Look at the photographs. With a partner, talk about how pollution has caused problems for these ocean animals.

How can we prevent these problems?

■ For more activities, go to Workbook 2 page 60.

## Litter survey

You are going to plan a litter survey.

1 Decide where in the school you will carry out your survey.

**Warning!** Never pick up litter with your hands. Check with your teacher and use a grabber and wear gloves.

2 Collect the litter from your chosen area.

3 Group the litter into different types. Use the table below to help you.

4 Count how much litter you find of each type. Record your results as tallies. Are there any patterns in your results?

| Type of litter | Number of items found |
|---|---|
| plastic | |
| paper | |
| cardboard | |

5 Design a poster to help to reduce the amount of litter in this area.

Check how much you know.
Try the puzzles on pages 62–63.

■ For more activities, go to Workbook 2 page 61.

## Be a scientist

Scientists collect information from surveys more than once and in more than one place.

▶ page 9

## Key idea

Not caring for our environment kills plants and animals, and pollutes water.

**1** Which of these things is living and which is non-living?

Write the words in the table.

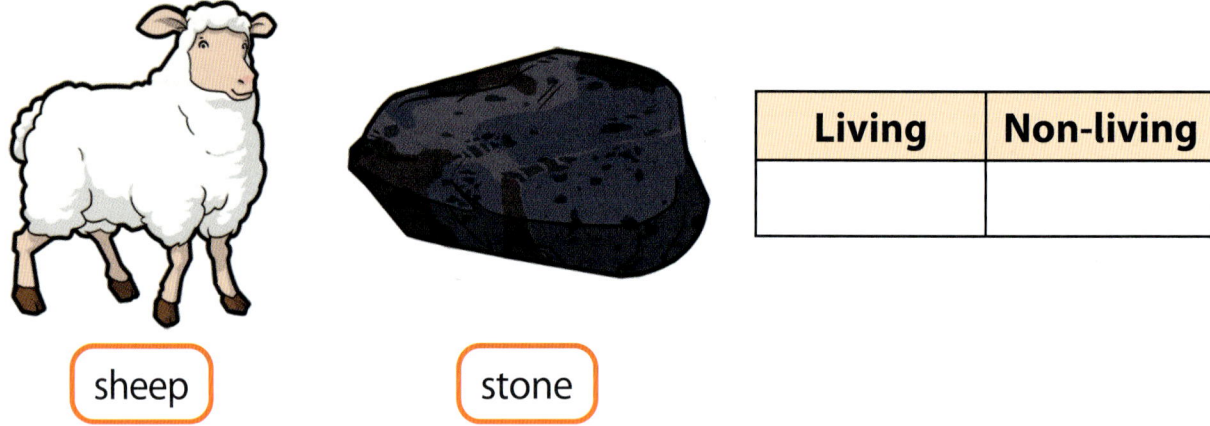

sheep

stone

| Living | Non-living |
|--------|------------|
|        |            |

**2** Draw a line from each plant or animal to the place where it lives.

desert

forest

Arctic

sea

chimpanzee

starfish

cactus

polar bear

■ For more activities, go to Workbook 2 page 62.

**3** Tick the photograph below that shows a micro-habitat.

**4** Which plant can live in the desert? Circle your answer.

**5** Circle the producer in the food chain.

Underline the herbivore.

Tick the carnivore.

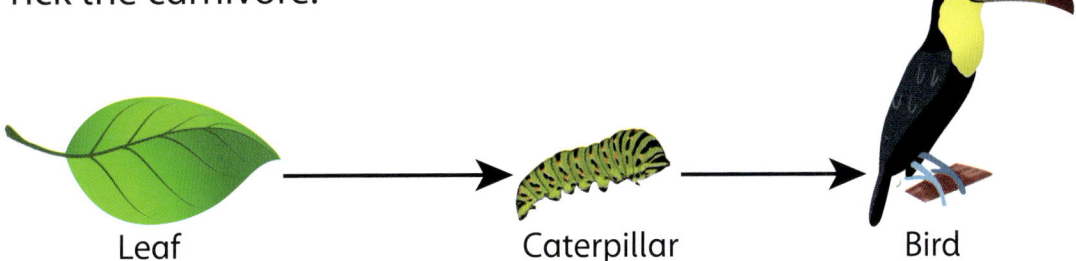

Leaf        Caterpillar        Bird

**6** Underline the examples of damage to environments that are caused by human activity.

> deforestation   earthquakes   hurricanes   litter
> plastic pollution   vehicle fumes   volcanoes

**7** Write down two ways that camels are adapted to live in the desert.

_____

_____

In this unit you will:

- find out about the different properties of materials
- explore why different materials are used for different purposes
- discover how the shapes of some materials can be changed
- compare and sort materials into groups.

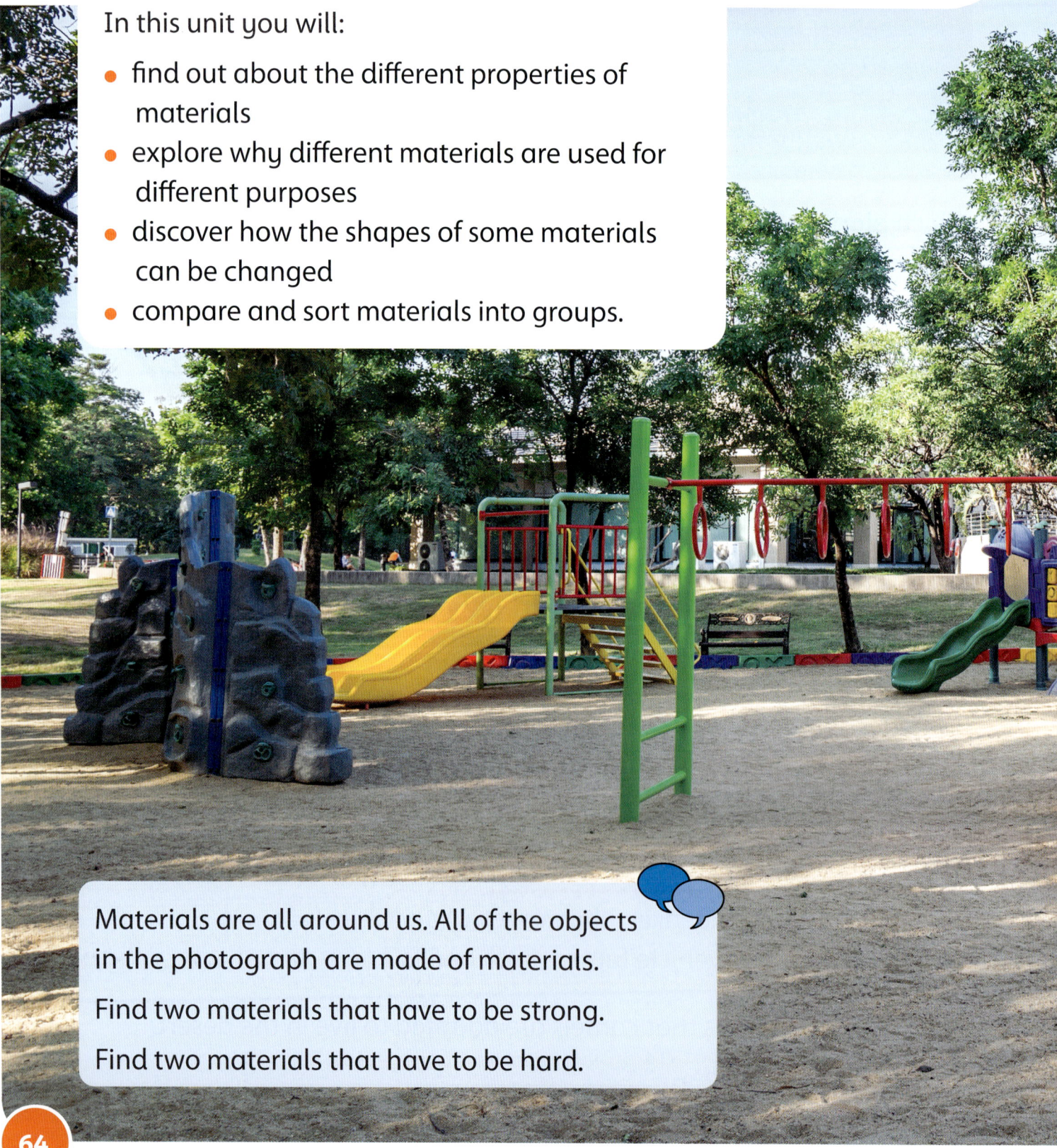

Materials are all around us. All of the objects in the photograph are made of materials.

Find two materials that have to be strong.

Find two materials that have to be hard.

absorbent   hard
human-made
material   natural   properties
soft   waterproof

**Science fact**

John Dunlop invented the first blow-up tyre made out of rubber. It was for his son's tricycle. He sold his invention and Dunlop tyres are now used all over the world.

These objects are made of glass.

Why is glass such a useful material for a jar?

Why aren't hammers made from glass?

■ For more activities, go to Workbook 2 pages 64–65.

# Materials around us

In this lesson you will find out that every material has properties.

**Key words**
hard/soft
materials
properties
rough/smooth

All the objects we use are made of materials. Even our bodies are made of different materials.

We can identify the different materials because they look and feel different. They have different properties.

**Think back**

Can you remember some of the words you used before about properties of materials?

How many useful materials can you see around you?

## What is in the box?

Work with a partner. You are going to investigate different properties. Use what you know about properties to help you to identify the mystery objects.

1 Place an object in the mystery box. Do not let your partner see the object.

2 Ask your partner to put their hand in the box. Ask them to try to identify the object and the material.

3 Take turns to put an object in the box.

4 Which properties did you use to identify objects?

5 Which objects were difficult to identify?
Which object was the easiest to identify?

66

■ For more activities, go to Workbook 2 page 66.

## Matching materials and properties

Can you match each material to its properties? Write the answers in your notebook. One has been done for you: a = 5.

**a** metal bowl

**b** cardboard box

**c** glass bottle

**d** rubber ball

**e** wool rug

**f** stone for building

**1** This material is hard and rough. It is heavy and strong.

**2** This material is see-through. It breaks easily.

**3** This material is soft and feels hairy.

**4** This material is not see-through. It is light and bends.

**5** This material is hard and shiny. It can be hammered into shape

**6** This material is soft and bouncy.

**Stretch zone**

Which senses did you use in each of these investigations?

**Key idea**

Materials have different properties that can be very useful.

■ For more activities, go to Workbook 2 page 67.

# Some materials can be used for a special job

In this lesson you will explore how the properties of different materials help them to do different jobs.

**Key words**
material
property

## Think back

Can you name two properties of each of these materials?

- rubber
- brick
- metal
- glass

What are the uses of metals in the photographs?

Discuss with a partner all the uses of metals you can think of.

Metals are very useful. People have used metals for thousands of years.

Some metals stay shiny for thousands of years.

Other metals react with water and air, and quickly change into dull materials. For example, iron can rust if it gets wet.

Rusty iron screws

■ For more activities, go to Workbook 2 page 68.

## Materials game

Play this materials game with a partner.

**Student A:** Say the name of an object.

**Student B:** Tell Student A the best material to make the object from.

> Metal is the best material for a spoon.

**Student A:** Ask Student B which is the worst material to make it from.

> Paper is the worst material for a spoon.

Continue until you have both chosen four objects each.

Look at these words. Decide which ones are properties of materials. Make a list of all the properties.

| | | |
|---|---|---|
| smooth | hard | |
| measure | strong | |
| see-through | happy | |
| metal | dull | rough |
| soft | shiny | bendy |
| heavy | copper | |
| observe | ball | paper |
| compare | | |

### Stretch zone

A gold car would last longer than a steel car. Use books or the internet to find out why we don't use gold to make cars.

### Key idea

Different materials have different properties. This makes them useful for different jobs.

■ For more activities, go to Workbook 2 page 69.

# Using the properties of materials

In this lesson you will continue to explore how the properties of different materials help them to do different jobs.

It is important to link the properties of a material to its uses.

## Materials in your school

We use materials for different things because of their properties.

1  Walk around the classroom or your school. Try to find different materials.

2  Note where you find each material.

3  Test each material to find out what properties it has.

### Stretch zone

Can you think of a better material to use for one object? Explain your answer.

## Key words
grippy
shiny
slippery

## Science fact

Many things are a mixture of two or more materials. Materials keep their properties in a mixture acting as before they were mixed. Soil and sea water are examples. Materials in a mixture can be separated again.

How will your group record your results?

Which properties will you look for?

■ For more activities, go to Workbook 2 page 70.

The materials in the boxes have different properties. Match each material to the sentence describing its properties. Tell your partner a use for each material.

glass    rubber    steel    chalk

**1** I am shiny, hard and very strong. I can be shaped to make many different objects.

**2** I am dull and soft. I am white and I am easy to see on black surfaces.

**3** I am hard and shiny. You can see through me. Be careful, I break easily.

**4** I am bendy and stretchy. I can have different colours.

## Slippery or grippy?

Shiny floors can be very slippery. This means that people might slip and fall over.

Think about the floors in your school. Which floors are shiny?

Do not run!

What might happen if shiny floors get wet?

Look at the sign at a swimming pool. How does this sign help you to be safe?

**Key idea**

Properties of materials make them suitable for different jobs.

■ For more activities, go to Workbook 2 page 71.

# Just right for the job

In this lesson you will continue to explore how the properties of different materials help them to do different jobs.

**Key words**
fabric
waterproof

Clean up water?

Hammer nails?

Make a mirror?

Make a bell?

Look at the photograph of paper towels.

Which job do paper towels do best?

Why are they not good for the other jobs?

Now you will investigate a group of materials called fabrics.

A fabric is a cloth made by weaving or knitting. Fabrics are used for carpets, curtains, clothes and coverings for furniture.

weaving

knitting

This person is selling lots of different fabrics

72

■ For more activities, go to Workbook 2 page 72.

waterproof? hardwearing? soft? strong? lightweight? bendy?

With a partner, discuss each of the properties in the picture. Are they important for the fabric for a T-shirt?

## Which fabric is best for a warm blanket?

Imagine you are choosing a fabric for a warm winter blanket.

1  How will you test your fabrics to find out:

- which are bendy?
- how strong they are?
- how soft they are?
- which is the warmest?

2  Carry out your tests.

3  Decide which is the best fabric for the blanket.

4  Imagine you work for a shop that sells this fabric. Design a presentation to explain why your fabric is the best.

### Stretch zone

Research the work of Charles Macintosh. Make a small poster about his work and what he invented.

### Key idea

A material must have just the right properties to do its job.

■ For more activities, go to Workbook 2 page 73.

# Sorting materials

In this lesson you will sort materials into groups.

**Key words**
cardboard
fabric
glass
metal
plastic
wood

**Think back**

Think about the different properties of materials you have learned about so far.

We can group materials into families or groups depending on their properties.

## Make a display of materials

You will need a large piece of card and lots of small objects.

You are going to sort the objects into groups and make a display.

You can sort the objects into different materials:

metal | wood | cardboard | plastic | glass | ceramic | fabric

Or you can sort the objects into different properties:

hard | soft | smooth | rough | bendy | rigid

heavy | light | absorbent | waterproof

see-through | opaque | shiny | dull

1 Draw lines on your card to divide it into squares. Draw one square for each material or property.

2 Write the name of a material or property in each square.

3 Put an example of each object in the correct square, or you could draw it.

■ For more activities, go to Workbook 2 page 74.

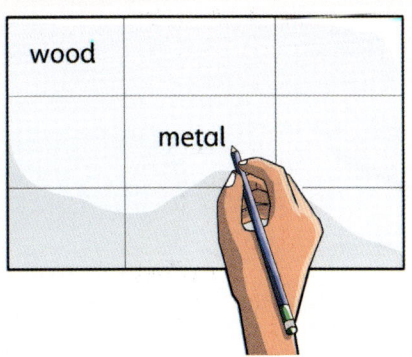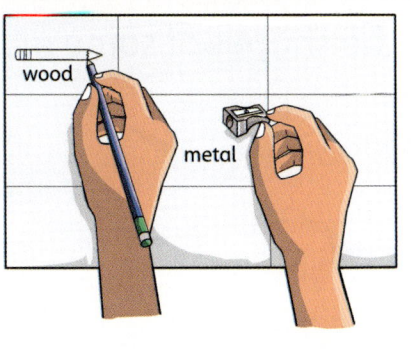

We can also split each group of materials into smaller groups. For example, there are hard woods and soft woods.

## Grouping materials

1 Write each of these material words onto post-it notes or cards.

> brick   cardboard   fabric   glass
> metal   plastic   wood

2 Move around the room and find objects made from these materials.

Stick the correct label onto each object.

3 Which material did you find the most? What properties of this material make it so useful?

### Be a scientist

Scientists often sort things into groups according to their properties or characteristics.

▶ page 10

**Stretch zone**

Research the different types of plastics. How many types are there? For example, think about a hard light switch and a soft plastic bottle.

### Key idea

We can sort and group materials depending on their properties.

4 Uses of Materials

75

# Changing the shapes of materials

In this lesson you will explore how we can change the shapes of some materials.

**Key words**
bend
shape
squash
stretch
twist

Discuss the ways you have changed the shape of something.

## Making a paper picture

1  Draw a picture in pencil on a large piece of card.

2  Take some pieces of coloured tissue paper and change the shape of them to fit into your picture.

3  Stick the coloured tissue paper to your picture.

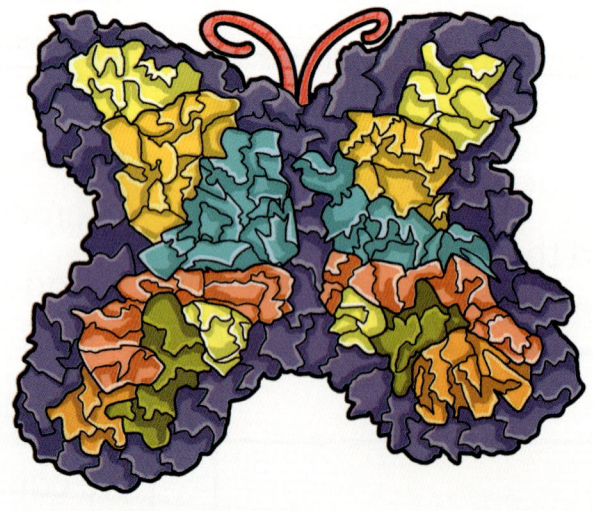

Talk to your partner about how you changed the shape of the paper. Use these three words:

• squash
• twist
• bend.

Bending, squashing and twisting a material will change its shape.

76

■ For more activities, go to Workbook 2 page 76.

Another way to change the shape of a material is to stretch it.

## Stretching materials

You are going to find out which materials stretch the most.

1 Choose your material. Measure how long it is.

2 Fix your material to the holder.

3 Hang the weights from it with a paperclip.

4 Measure and write down how many centimetres the material has stretched.

5 Repeat with other materials.

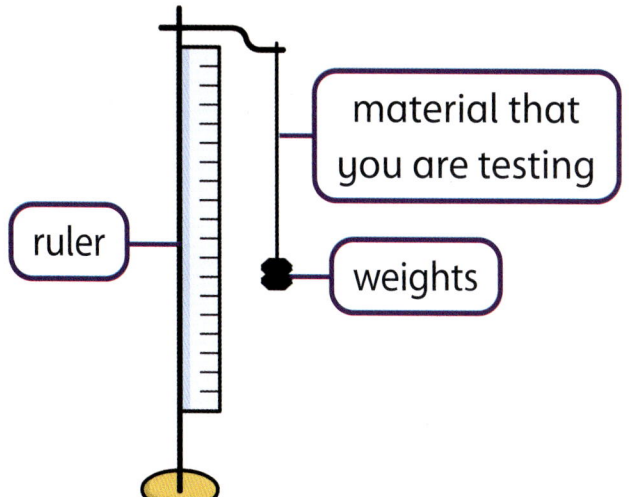

ruler

material that you are testing

weights

6 Write down in your notebook which material stretched the most.

**Be a scientist**

A scientist uses the same length of material in every test.

Why?

▶ page 8

Discuss why is it important that some materials stretch and others do not.

**4 Uses of Materials**

**Stretch zone**

Imagine a world where bricks stretch. Write a short story about what problems people would have.

## Key idea

We can make things change shape by squashing, twisting, bending and stretching them.

77

■ For more activities, go to Workbook 2 page 77.

# Heating materials

In this lesson you will explore the way some materials change when they are heated.

**Key words**

change
heat
ice
melt
steam
water droplets

**Think back**

Think about the four ways that you have changed materials by twisting, bending, squashing or stretching.

**a**

**b**

Look at the photographs. Which object is changed by bending?

Which is changed by twisting?

Materials can also be changed by heating them. Heating something means to make it hotter.

**Science fact**

Ice changes into water if the temperature is above 0 degrees Celsius. How warm is your classroom?

Talk about what you predict will happen to these ice cubes if left on a sunny windowsill.
You could use an ice cube to test your ideas.

■ For more activities, go to Workbook 2 page 78.

## Making clay bowls

Your teacher will give you some clay. You are going to make two bowls.

1  Split your clay into two equal pieces.

2  Make each piece into a small bowl of the same size and shape.

3  Put one bowl into a cool, damp place. Your teacher will heat the other one so that it gets very hot.

4  Predict what you think will happen to each bowl.

5  Once you get your bowls back look to see how they have changed. Were your predictions correct? How has the heated bowl changed?

## Be a scientist

Scientists carefully observe and record what happens during an investigation. You could take photographs of your pots before and after heating.

▶ page 10

What is happening in the diagram? Use the words heat, water, steam and water droplets.

## Key idea

We can change materials by heating them.

■ For more activities, go to Workbook 2 page 79.

# Changing foods by heating them

In this lesson you will explore the way some foods change when they are heated.

**Key words**
baking
melt

Lots of other materials change when they are heated. This includes foods.

## How foods change when they are heated

You are going to do some baking. Your teacher will give you the ingredients.

125g butter

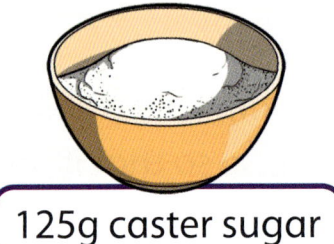

125g caster sugar

2 medium eggs

2 tablespoons of milk

125g self-raising flour

1  Mix together the butter and sugar in a bowl until the mixture feels light and fluffy.

2  Add the eggs, flour and milk to the bowl and mix it all together until the mixture feels smooth.

**Warning!** Your teacher will put the cakes in the oven, then take them out when they are ready.

3  Put your mixture into paper cases and bake the cakes in the centre of the oven.

■ For more activities, go to Workbook 2 page 80.

How does heating change the cakes?

Do the cakes look different from the ingredients you mixed together?

Do the cakes smell and feel different?

## Think back

Think back to the ice cube left in the Sun. The ice cube melted into water.

You are going to investigate ways to make ice melt more quickly.

## Investigating melting

You will be given five ice cubes and a timer.

1  Place the ice cubes in different parts of the room.

2  Predict which ice cube will melt first.

Predict which ice cube will melt last and explain why.

3  Carry out your investigation. Use a timer to find out if your predictions were correct.

4  Write down your times in a table.

5  Were your predictions correct?

## Science fact

Making steel is a bit like baking a cake. Ingredients are added together and heated. Your cake was baked at 200°C. The steel is 'baked' at 1200°C!

## Key idea

We can change foods by heating them.

■ For more activities, go to Workbook 2 page 81.

# Cooling materials

In this lesson you will explore the way some materials change when they are cooled.

**Key words**
cool
freeze
liquid
solid

You have learned that materials can change if you heat them. Materials can also change when they are cooled.

Cream is a liquid. When cream is cooled it changes.

Discuss how the cream changes when it is cooled. What do we call the changed cream?

 **Making ice pops**

You are going to make an ice pop. Your teacher will give you a mould and some milk or juice.

1  Carefully pour the liquid into the mould.

2  Be careful not to spill any of the liquid. Do not fill the mould all the way to the top.

3  Put a wooden or plastic stick into the liquid.

4  Place the filled mould into a freezer and leave for a day.

5  Predict what will happen to your liquid milk or juice.

6  When you get your mould back describe what you see.

Does the milk or juice look different?
Does the milk or juice feel different?

 **Be a scientist**

Scientists always plan an investigation before they start. This helps them to know what equipment and materials they need.

▶ page 8

■ For more activities, go to Workbook 2 page 82.

## Design an advert

1   Design an advert for your ice lolly.

Include a large bright picture.

Give your lolly a name.

2   Explain how your ice lolly was made.

Use the words from the box below and the instructions for making ice pops on page 82.

> freeze    liquid    solid

Use the words from the box below and the instructions for making ice pops on page 82.

### Science fact

Scientists have discovered what causes brain freeze. When the cold touches the roof of your mouth, it affects the blood flowing to the brain.

## Stretch zone

What do you predict will happen to an ice pop if you observe it for too long? Could you make it back into an ice pop again? Write instructions of how you can do this.

## Key idea

We can change materials by cooling them.

 For more activities, go to Workbook 2 page 83.

# Dissolving materials in water

In this lesson you will discover that some materials dissolve in water.

**Key words**
dissolve
material
soluble/insoluble

If a material spreads out in water so that we cannot see it, we say it has dissolved.

The material may colour the water but we cannot see the pieces of the material.

If a material does not spread out in water we say it has not dissolved.

A material that dissolves in water is called soluble. When a material dissolves in water it forms a type of mixture called a solution.

A material that does not dissolve in water is called insoluble.

With a partner, talk about the photograph of the cup of tea.

What is happening to the tea bag?

What is happening to the water?

Look at the photographs below. They show materials before and after they have been put into water.

Which material has dissolved to make a solution and which has not dissolved?

■ For more activities, go to Workbook 2 page 84.

## Dissolving

You will be given different materials to test if they dissolve in water.

1  Add some water to your beakers so they are all half full.

2  Predict whether the materials will dissolve or not.

3  Add a spoonful of the first material to your first beaker. Stir.

4  What happens? Was your prediction right?

5  Test all of the materials in the same way.

6  Design a results table to record your findings. Are there any patterns in your results?

We know that sea water is salty but the sand on the beach does not dissolve into the sea.

 **Stretch zone**

How could you show that when things dissolve they do not disappear?

### Be a scientist

Scientists only change one thing in an investigation. You are changing the material. Why should you use the same volume of water every time?

▶ page 8

With a partner, talk about why it is important that some materials dissolve and others do not.

### Key idea

Some materials dissolve in water and others do not.

85

# Natural or not natural?

In this lesson you will learn that some materials occur naturally and others are human-made.

**Key words**
human-made
natural

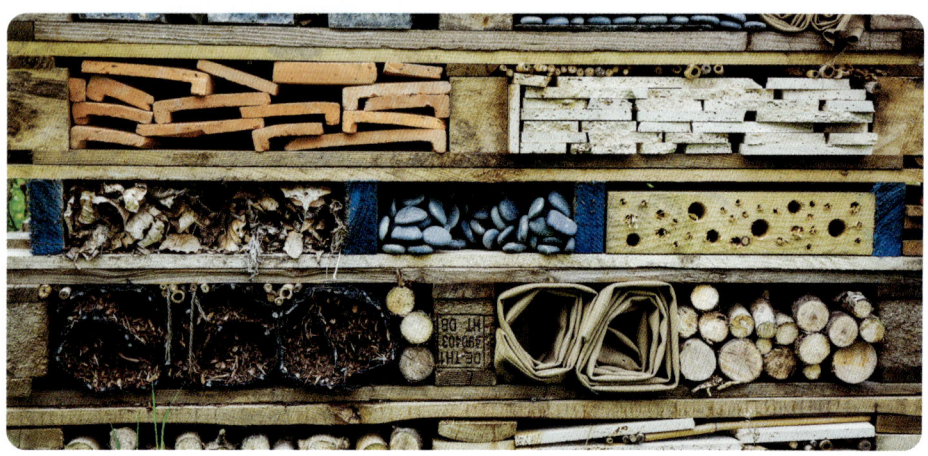

A display of natural materials

Materials found in nature are called natural materials. Natural materials include wood, clay, stone, shell and sand.

With a partner or small group, talk about the materials in the photograph.

How many can you name?

Can you think of some uses of these materials?

Point to how each natural material is used. One has been done for you.

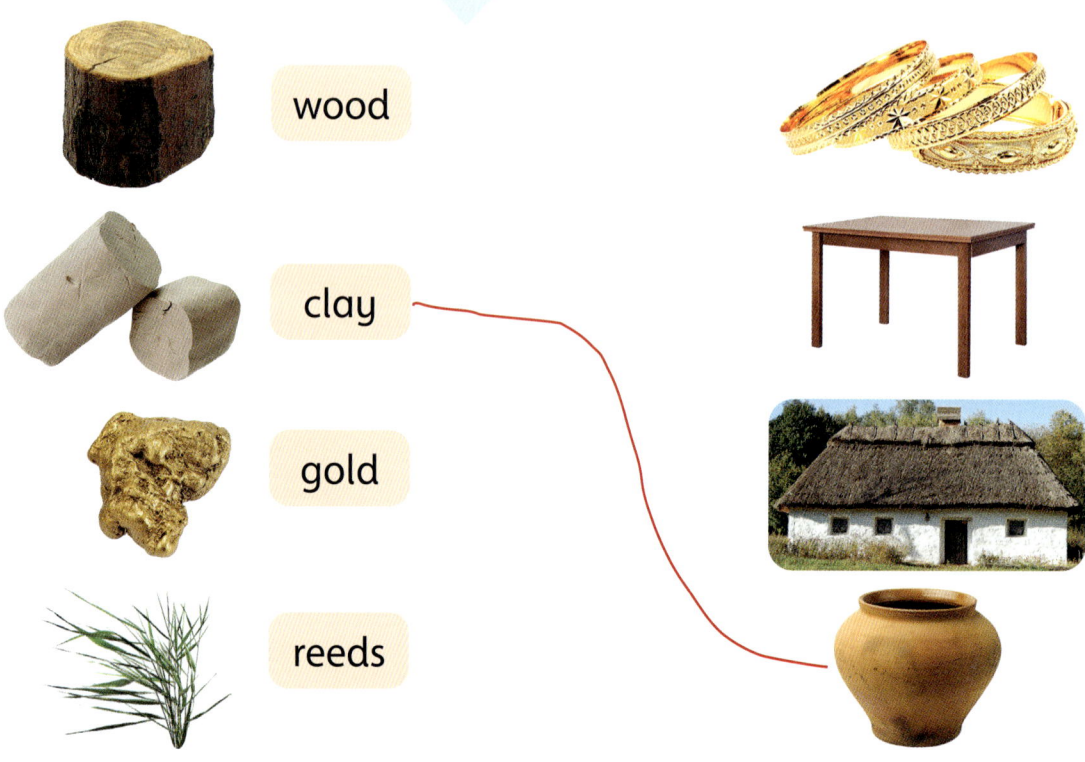

wood

clay

gold

reeds

■ For more activities, go to Workbook 2 page 86.

Look at the photograph of the building.

Write down all the human-made materials in your notebook.

Materials that people make are not natural. They are human-made. Human-made materials include glass, plastic, concrete and steel.

Many buildings, roads, dams and bridges are made from bricks and concrete.

## Sorting objects – natural or human-made?

You are going to sort objects into groups of natural and human-made.

1  Look at the objects your teacher has given you.

2  Decide which materials the objects are made from.

3  Draw two large sorting circles on your playground with chalk and label them as in the picture.

4  Place the objects into the correct sorting circle.

5  Take a photograph and use it to help you make a display of natural and human-made materials.

### Stretch zone

Imagine what your school would look like if there were no human-made materials.

Write a short story about how it would be different and what you would miss.

Check how much you know.
Try the puzzles on pages 88–89.

### Key idea

We use materials that are found in nature and some that are made by people.

■ For more activities, go to Workbook 2 page 87.

# What have I learned about uses of materials?

1 Draw a line between the two words in the clouds and any objects in the photographs that have that property.

hard      soft

2 We can sort objects into groups according to their properties or the material they are made of.

a Circle the words that are properties.

b Underline the words that are materials.

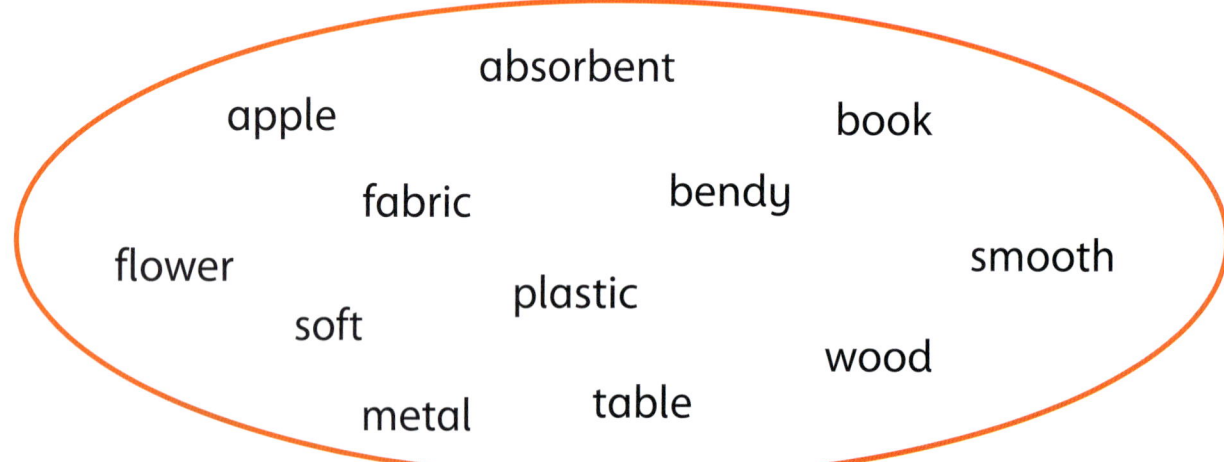

absorbent

apple          book

fabric     bendy

flower                    smooth

plastic

soft

wood

metal     table

3 How does water change when it is heated? Circle the correct word.

When heated, water becomes:  ice   steam   water

■ For more activities, go to Workbook 2 page 88.

4 The pictures show how the shape of a material is changed. Label each picture using the words from the word box.

> bending   squashing   stretching   twisting

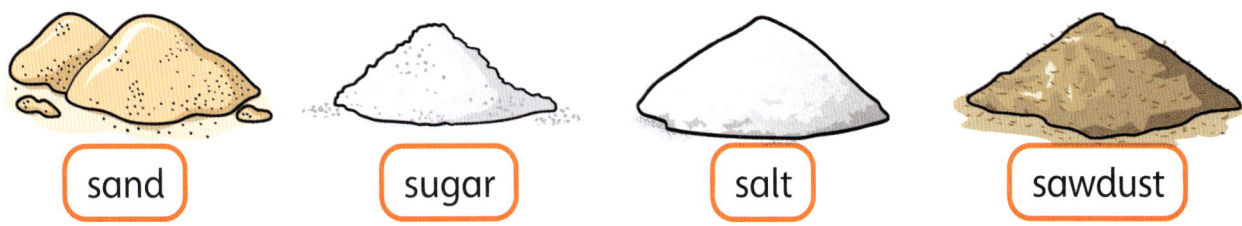

|  |  |  |  |
|---|---|---|---|

5 Circle the substances that will dissolve in water.

sand        sugar        salt        sawdust

6 A student puts four ice cubes at different places in the classroom. They measure the time each one takes to melt. The table shows the results.

| Place | Time taken to melt |
|---|---|
| In a cupboard | 14 minutes |
| On a desk | 13 minutes |
| Windowsill | 10 minutes |
| On the floor | 15 minutes |

a In which place did the ice cube melt first?

_____

b In which place did the ice cube melt last?

_____

■ For more activities, go to Workbook 2 page 89.

In this unit you will:

- explore how the Sun appears to move during the day
- show how the spin of the Earth gives day and night
- describe the change in the shape and position of the Moon
- investigate how shadows change.

Look at the photograph of the Sun.

With a partner, talk about the words you can use to describe it.

**Science fact**

Did you know that the Sun is actually a star? It is the biggest star in the sky. The Sun is bigger than one million Earths. Can you imagine that? The Sun is our main source of light.

dark   day   Earth
light   Moon   night
shadow   Sun

Look at the photograph of the night sky. What do you see at night that you don't see in the daytime?

**Stretch zone**

What happens to the Sun at night?

■ For more activities, go to Workbook 2 pages 90-91.

# The Sun appears to move in the sky

In this lesson you will explore how the Sun seems to move during the day.

**Key words**

axis     spin
Earth    Sun
light

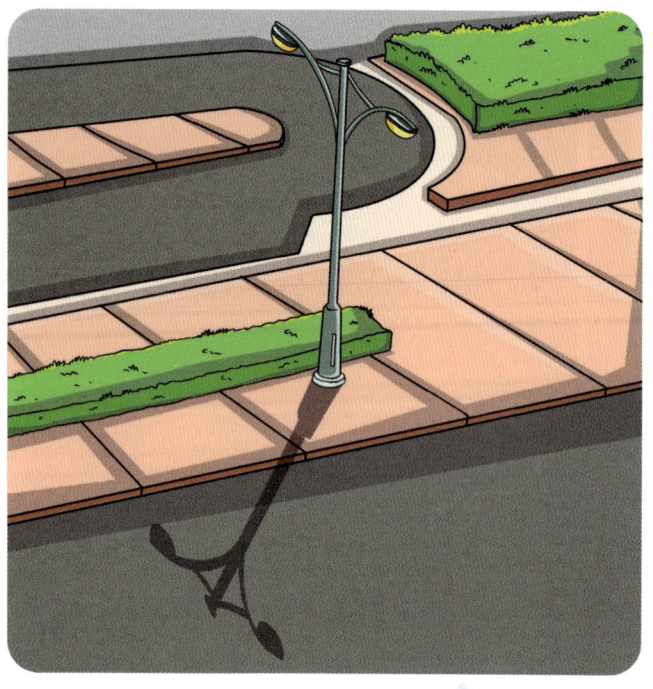

Look at the pictures. With a partner talk about how they are different. Decide on a reason why this happens.

The Sun gives us light. It is the main source of light on Earth.

The Sun appears over the horizon in the morning. It then seems to move higher across the sky.

From midday the Sun seems to move lower and lower until it disappears below the horizon.

Does the Sun actually move across the sky?

Talk about your ideas as a class.

**Warning!**
Do not look directly at the Sun – it will hurt your eyes. In summer and at midday the Sun is very powerful. It can burn our skin even though it is far away.

92

■ For more activities, go to Workbook 2 page 92.

People say that the Sun rises in the morning and goes down at night. This is confusing because it makes us think that the Sun moves across the sky.

This is not true. It is actually the Earth that is moving.

Look at the diagram of the Earth. The axis of the Earth is an imaginary line through the Earth from the North Pole to the South Pole.

The Earth spins on its axis. This takes 24 hours.

axis

## Following the Sun

You teacher will stick some paper onto the window where the Sun shines through early in the morning.

1   Draw what the window looks like in the morning.

    More paper will be added every hour to cover where the Sun shines through.

2   Draw what the window looks like at midday.

3   Draw what the window looks like at the end of the school day.

What shape did the Sun make as it appeared to move across the sky?

When was the Sun at its lowest? When was the Sun at its highest?

### Key idea

As the Earth spins, the Sun looks as if it moves across the sky. But it does not.

■ For more activities, go to Workbook 2 page 93.

# Tracking the Sun and Moon in the sky

In this lesson you will track how the Sun and Moon appear in the sky.

### Key words
compass
east/west
midday
Moon
morning/evening
orbit
phase

### The Sun survey

1 Use a pin to make a small hole in the middle of a piece of paper.

2 Go outside and hold the paper up.
Without looking directly at the Sun, point the hole at the Sun.

3 Move another piece of paper around until you see an image of the Sun on it.

4 Move the paper backwards and forwards until the image of the Sun has sharp edges.

5 Use a compass to find out the direction of the Sun. Record the direction in your notebook.

6 Go back to the same place at different times of the day.
Predict where you expect to see the Sun.

7 Use your pieces of paper to observe the image of the Sun and use a compass to record its direction at the different times of day.

Were your predictions correct?
At what time was the Sun above you?

**Warning!**
Do not look directly at the Sun.

**Be a scientist**

Scientists work very carefully to keep themselves and other people safe. They use the correct equipment.

▶ page 9

94

■ For more activities, go to Workbook 2 page 94.

The Sun appears to change during the day.

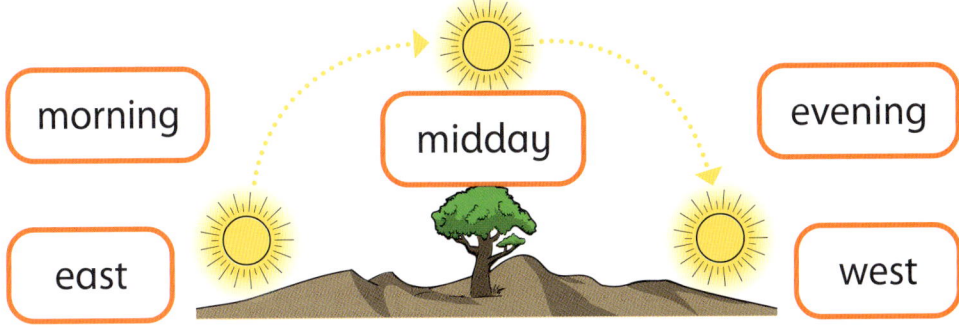

morning | midday | evening

east | west

## Phases of the Moon

The Moon moves around the Earth once every month. This is called its orbit. The Moon looks a different shape, or phase, during a month. Its shape depends on how much of it is catching the light from the Sun during its orbit.

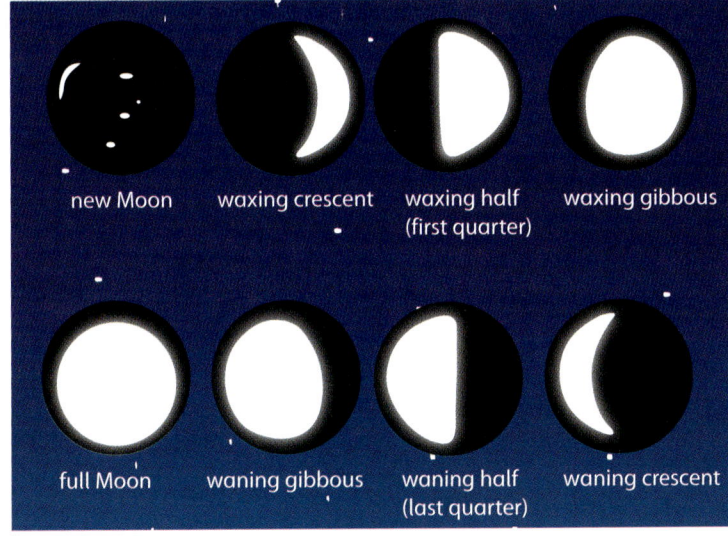

new Moon | waxing crescent | waxing half (first quarter) | waxing gibbous

full Moon | waning gibbous | waning half (last quarter) | waning crescent

## Modelling phases of the Moon

1   Use biscuits or modelling clay to make the phases of the Moon during a Month.

2   Write a label for each phase.

3   Display your model in your classroom.

Look at the diagram. What does it tell you about how the Sun appears to move across the sky? Share your ideas with a partner about whether the Sun is really moving.

## Key ideas

- As the Earth spins, the Sun appears to rise in the east and set in the west.

- The orbit of the Moon around the Sun makes the Moon a different shape throughout a month.

**5 Day and Night**

■ For more activities, go to Workbook 2 page 95.

# Shadows change during the day

In this lesson you will discover how shadows change length and move during the day.

**Key words**
compass
dark
direction
length
shadow

## Think back

When you did your survey of the Sun, did you notice your own shadow?

A shadow is formed when an object blocks the light.

A shadow is the dark shape of the object that is blocking the light.

In a dark room, we can make shadows by using a light source such as a torch.

## Exploring hand shadows

1  Hold a torch very close to your partner's hand.

2  Ask your partner to form a shape with their hands.

3  Now move the torch up and down and further away from your partner's hand.

Discuss what happens to the length of the shadows and the direction they point each time.

■ For more activities, go to Workbook 2 page 96.

## Exploring shadows

1   Set up a shadow stick as shown in the drawing.

2   Use a ruler to measure the length of the shadow. Use a compass to measure its direction.

3   Predict the direction and length of the shadow in one hour, two hours, three hours and four hours.

4   Measure the shadow every hour to test your predictions.

5   Design a short presentation of your results to tell the class about your investigation.

You could make a poster or a computer presentation, or give a short talk.

## Science fact

When the Moon passes between the Sun and the Earth it casts a huge shadow. This is called an eclipse. It makes parts of the Earth very dark in the daytime.

## Key idea

As the Sun appears to move across the sky it makes shadows change shape and direction.

■ For more activities, go to Workbook 2 page 97.

# The Earth is spinning

In this lesson you will show how the spin of the Earth produces day and night.

**Key words**
day/night
spin

**Think back**

How long does it take the Earth to spin once on its axis?

The Sun lights up the part of the Earth that is facing the Sun.

Look at the picture of the Earth and the Sun with your partner.

- Point to the part of the Earth where it is day.
- Point to the part of the Earth where it is night.

Can you agree on your answers to the next questions?

- After 12 hours, what will happen to the parts that are in day?
- What will happen to the parts that are in night?

**98**

■ For more activities, go to Workbook 2 page 98.

## Spinning Earth

We can model the Earth and the Sun with a torch and a ball.

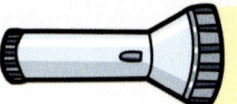 

The ball can be the Earth. What is the torch modelling?

1  Work with a partner. Choose to be either the Sun or the Earth.

2  Put a mark on the ball to show where we are on the Earth.

3  Hold the torch very, very still! Hold the Earth in front of the torch and slowly turn it.

4  Which part of the model is showing night and which part is showing day?

5  What happens to the place on the ball where you made a mark?

6  How does this explain how the spinning of the Earth gives day and night?

### Stretch zone

It takes 365 and a quarter days for the Earth to move around the Sun. What else do you know that has 365 days? What happens to all of the quarter days?

### Key ideas

- The Earth takes 24 hours (one day) to spin around.
- We get day and night because the Earth spins and different parts of Earth are facing towards the Sun or away from it.

Check how much you know.
Try the puzzles on pages 100–101.

### Science fact

Some countries have almost 24 hours of sunlight in the summer and almost 24 hours of darkness in the winter.

**5 Day and Night**

99

■ For more activities, go to Workbook 2 page 99.

# What have I learned about day and night?

**1** Underline the correct answer.

**a** When your country is pointing away from the Sun, it is:     day   night

**b** When your country is pointing towards the Sun, it is:     day   night

**2** Circle the time of the day that the Sun appears highest in the sky.

early morning    evening    mid-afternoon    midday    mid-morning

**3** Write the labels of the phases of the Moon. Use the words in the word box.

full Moon     new Moon     waxing crescent Moon

■ For more activities, go to Workbook 2 page 100.

**4** How long does it take the Earth to spin around on its axis?

<br>

**5** Look at the picture. Draw in where you think the tree shadow will be:

Morning shadow

**a** at midday

**b** in the evening

Think about the direction and length of the shadows.

**6** Tick the two statements that are correct:

Shadows always point east. ☐

Shadows change in length and direction during the day. ☐

Shadows are longest at midday. ☐

Shadows are shortest at midday. ☐

**7** Write the words in the correct post-it on the picture, to show where the Sun is at different times of the day.

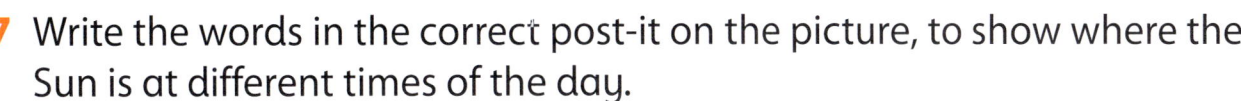

evening  midday  morning

■ For more activities, go to Workbook 2 page 101.

# Glossary

**absorbent**

**adapted**

**adult**

**bulb**

**dark**

**day**

**diet**

**Earth**

**environment**

**exercise**

**food chain**

**germination**

**grow**

**habitat**

**hard**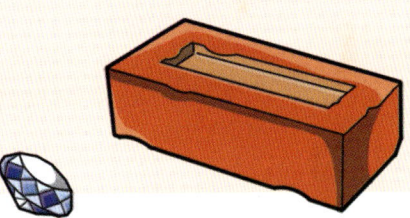

**human-made**

**hygiene**

**light**

**living**

**material**

**micro-habitat**

**minibeast**

**Moon**

**movement**

**natural**

**night**

**non-living**

**offspring**

**parent**

**plant**

**pollution**

**properties**

**seed**

**shadow**

**soft**

**Sun**

**teenager**

**temperature**

**toddler**

**water**

**waterproof**